Contents

Introduction

The last decade saw collective insanity sprouting around bitcoin and other cryptocurrencies. Still, it is hard to predict whether crypto is a good investment or a passing storm. While some investors remain skeptical about crypto investing, statistics show that bitcoin investments are gaining popularity among institutional and retail traders alike. Here is the most awaited eBook for you to learn all about Bitcoin. The book has five chapters that will take you from the very basics of cryptocurrency investment to advanced cryptocurrency trading. Go word by word if you are just starting up with the cryptocurrencies, and you will have all that you need to know about it. However, you can skip directly to "Chapter 5" if you want to get started and trade Bitcoin right away.

Bitcoin is climbing at an all-time high after it surpassed $19,000 first since 2017. It climbed more than 40% in November 2020 and saw its price double in 2020. Online investing apps are seeing first-time buyers flooding them with orders. Also, payment apps like PayPal and digital wallet providers are seeing a jump in crypto sales. Here are a few key statistics regarding the growing interest in Bitcoin trading:

- As per a Bloomberg article, Bitcoin trades per day were 81% higher than the rest of 2020 combined. Additionally, as per Google Trends, Google searches for the "Coinbase" crypto exchange were at an all-time high in at least a year.
- Square Inc., a San Francisco-based payments app developer, said that its customers purchased over $1.6 billion worth of Bitcoin through its Cash App in the third quarter of 2020, which was 83% high from the prior period.
- According to data from CoinGecko, after the Reserve Bank of India's (RBI), April 2018 circular was struck down by the Supreme Court, top Indian exchanges saw daily crypto trading volumes grow by over 500% since March 2020. The Supreme Court removed the Bank's restrictions on working with crypto firms, and the COVID pandemic encouraged retail traders to start investing in cryptocurrencies.

Thus, both as an investment and a currency, Bitcoin's overwhelming performance is attracting institutional and retail investors alike. It is an investment tool that provides investors with many benefits over traditional investments. But what is crypto? How does Bitcoin trade? What are the ways to make a profit when trading with Bitcoin? Will it even exist in ten years? We look at all these answers in the subsequent chapters.

CHAPTER 1

The Fundamentals Of Cryptocurrency Every Investor Must Know And Understand

Introduction

Adopting blockchain-based solutions is an increasing trend across enterprises, with financial services, non-food manufacturing industries, and media and communications leading the movement. The number of organizations that are intensifying blockchain integration is increasing. For leveraging the crypto technology for growth, it is crucial to understand the basics of cryptocurrency. 2020 may have been a troublesome year for many, but it was an excellent time for cryptocurrencies. The last few years have witnessed rapid adoption from influential organizations like PayPal, allowing users to buy, hold, and sell Bitcoin. Bitcoin has made an all-time high of over $40,000, bringing its market cap to over $642 billion, and interest has surged higher than ever before.

The advantages of cryptocurrency and Blockchain can go far beyond the small and medium sellers on e-commerce platforms. Banks around the world are recognizing the importance of using a faster and better-secured currency exchange platform. They are substituting the less secure payment processes with blockchain-based methods. Once the general public approves these methods, banks can incorporate Blockchain as a viable payment process tool. Cryptocurrencies have several use-cases and profit-making opportunities, but a user must be well informed to take full advantage of the boom.

1.1 What is a Cryptocurrency?

A Cryptocurrency like Bitcoin or Ethereum is a digital medium of exchange or store of value. Cryptocurrencies are purely digital and use various encryption mechanisms to verify fund transfers between holders and create monetary units. A Cryptocurrency is one of many kinds of digital currencies that allows the holder to purchase services and goods from merchants that accept it. At their core, cryptocurrencies use Blockchain, a decentralized ledger technology that provides privacy and security. The exact mechanism used in different cryptocurrencies may differ, giving some advantages over others that can lead to a corresponding edge in market value.

1.2 How Does Cryptocurrency Works?

Cryptocurrency transactions involve the usage of wallets. It is software that will assist the sender in completing the transaction. While transferring the virtual currency from one destination to another, a unique set of keystrokes is created. Whoever is the owner of that set of keystrokes becomes the owner of the amount.

Cryptocurrency is a decentralized system that has no government oversight, and the users are their banks. It gives them complete control over their money and transactions. However, being virtual cash, no system can create reversibility if a transaction has gone wrong. To understand cryptocurrency, we need to differentiate between cryptocurrency and fiat money.

1.3 The Difference Between Crypto And Fiat Currency

The fact that more people are transacting through electronic money affirms suggestions that crypto can become the future currency. However, it will take a while before it finds its way through to the mainstream sectors because of the strong opposition from regulators worldwide. Although the world is moving towards a cashless economy, few people know the difference between fiat and cryptocurrency. The following table will help put things in perspective.

Fiat Currency	Cryptocurrency
Fiat currencies have remained stable for most parts of their existence, and economies worldwide rely on them to continue trade and commerce.	By contrast, cryptocurrencies are based on a type of Decentralized Ledger Technology (DLT) known as Blockchain. Unlike fiat money, their system of creation is decentralized and based on cryptography and algorithms.
Countries keep control over the value of fiat money using monetary policies to keep the supply and demand for cash steady. Because of this, it is used as a store of value and facilitates business.	A blockchain refers to a digital ledger where transactions are recorded, duplicated, and distributed across several computer systems liked to the Blockchain.
Also, banks and financial institutions have a large role in fiat money. With activities such as lending and interest rates, they participate in the credit supply to keep the economy's engines running.	A block in a blockchain has records of transactions. Every time there is a transaction, the record gets added to the ledger of the participants.
Fiat money is the money issued by a government, which can be both digital or physical. It is held by banks and regulated by a central authority.	Blockchain is not limited to cryptocurrency. It has immense usage in other fields, like logistics, medical transcription, supply chain management, etc.

1.3.1 Disadvantages of Fiat Money

Fiat money has the following drawbacks.

- It is stable because of decisive governmental actions. However, currencies tend to get influenced by external events, and black swan situations can destabilize the supply and demand flow. F or i nstance, g overnment m ismanagement can have a domino effect on fiat, which causes it to become unstable. One example is Zimbabwe, where inflation got so high that it cost trillions of their currency to buy necessary survival items, making the currency worthless.

- Such concerns are valid when governments take action, such as continually decreasing interest rates and doing quantitative easing. Fiat currencies' value reduces over time as they are inflationary and thus harm the weaker sections of society who cannot afford to invest heavily in real-estate or equity. At the same time, the super-rich benefit since they have to pay less in debt. Fiat thus leads to inequality.

- Greater Government control leads to an overbearing bureaucracy. This overarching authority can hamper the privacy and freedom of individuals. Greater digitization means that governments can access a vast collection of data detailing all private transactions in an individual's life.

- Fiat also has high transaction costs for international fund transfers, and organizations such as Visa and Mastercard charge fees to allow purchases.

- Fiat money, being a physical entity, is also exposed to counterfeiting. Numerous successful counterfeit attempts have led governments to create tighter controls over monetary transactions. It has increased the overall cost of maintaining cash, which does have a domino effect on consumers.

1.3.2 Advantages And Disadvantages Of Cryptocurrencies

These disadvantages of fiat money have led to the development and rapid adoption of cryptocurrencies. While cryptocurrencies may only seem to have pros, there are cons too. The table lists the ***advantages and disadvantages of cryptocurrencies***.

Advantages	Disadvantages
Decentralization means that the presence of an overarching central authority is absent, and people can make use of their finance without censorship or breaches of privacy.	For a start, cryptocurrency may not be everyone's cup of tea. It is quite challenging to understand the entire process, and since there is no higher authority involved, one may not be able to look for help.
The Blockchain is a peer to peer network. Every wallet has a key that is known to the owner of the wallet only. The dependence on algorithms to generate currencies leads to faster decision making and quicker turnaround times. Also, several cryptocurrencies make transactions a smoother process with a negligible fee.	One of the pitfalls of any transaction is reversing it. While it is relatively easy in a bank, it becomes a headache when it comes to cryptocurrencies. It is also a primary reason why many have shunned virtual currency. As it stands now, there cannot be a reversible transaction in cryptocurrency.

- Being virtual, decentralized, and fortified with cryptographic algorithms, there is very little chance of forgery. Physical notes may turn out to be fake, but a Cryptocurrency transaction is genuine. For someone to break into the system and manipulate information would require almost 51% of the system's total computing capacity, which by itself is an impossibility.
- Cryptocurrencies, by their very nature, is a decentralized financial system. And because it does lack any administrative oversight, the chances of volatility increases manifold. There is no controlling factor, and investors are always at the mercy of the ecosystem's whims and fancy.

1.4 Cryptocurrencies: What Makes It A Better Choice Over Fiat?

While both types of currencies have their attributes, the future belongs to virtual cash. That is because of the privacy and security it provides to its users. Also, the absence of an institutionalized oversight helps it become a peer to peer network with no transaction fee and unnecessary paperwork. Here are a few reasons why the cryptocurrency is a better choice over fiat:

- ***Instant Settlement:*** This is attributed to the technology on which cryptocurrency is based, i.e., Blockchain. It is easy to use, and all one would need is a smartphone and an internet connection.
- ***Accessibility:*** Over a while, banks have become elitist, which has left many people unable to access a bank. They have been left out since they did not qualify for banking services. With cryptocurrency, there are no such qualification criteria. Anybody can be a part of the peer to peer network and become their bank.

- ***Security:*** One of the essential features of the Blockchain is its distributed ledger architecture. It creates numerous nodes or ledgers, each having records, albeit in encrypted form, of the transactions' owners. Because the entire ecosystem is decentralized, the virtual currency is owned by the individual. Also, the whole transaction is encrypted. It makes it challenging for hackers to break in and cause harm. The cash is then stored in digital wallets or crypto-wallets, the key of which is only with the owner of the wallet. Every key is a unique set of numbers.

- ***No Manipulation:*** A physical $10 bill can be copied and a fake created. It can also be torn, soiled, destroyed, etc. A virtual currency with no defined physical attributes will have no such challenges.

1.5 Different Types Of Cryptocurrencies

There are three major types of cryptocurrencies – Bitcoins, Altcoins, and Tokens. Bitcoin was the first currency that embraced the blockchain concept. After Bitcoin, there were several blockchains created, known as the altcoins. NEO, Cardano, and Litecoin are examples of altcoins. The third primary type of cryptocurrency is tokens. Its examples include Civic (CVC), WePower (WPR), and BitDegree (BDG).

- **Bitcoin:** Bitcoin introduced blockchain technology and allowed users to receive and send Bitcoins without using a third party. It is a digital currency that users can send to other people for a product, services, or as a gift. It is similar to the money we have in our bank accounts (EUR, USD, etc.), but it is digital. Every Bitcoin transaction happens between the users directly through a peer-to-peer network, possibly because of the blockchain technology.

- **Altcoins:** Next type of cryptocurrency is altcoins. There are thousands of altcoins in existence, most of which are minor variations or alternate versions of Bitcoins. While it may be why they get the name altcoins, not all altcoins are the alternate versions of Bitcoins. Some altcoins have different purposes and are different from Bitcoins.

- **Tokens:** The third type of cryptocurrency is a token, and they are the most interesting ones. The key distinguishing feature of tokens is that they don't have their personal Blockchain, unlike Bitcoins and altcoins. The dApps (decentralized applications) built on blockchains like NEO and Ethereum can transact using tokens. These apps use tokens because they are made to execute smart contracts.

Final Words

Bitcoin and cryptocurrencies have time and again proved that they are here to stay, no matter what. Every time the world thought the bitcoin is dead, it has risen to the occasion, even stronger than before. Investors are increasingly considering blockchain technology as the future of money. A contributing factor in this trend is the increase in the institutional investment from mainstream institutions and legacy influencers. This change has benefitted not only the Bitcoin but the broader crypto industry. This chapter discussed the basics of cryptocurrency, which will help investors know why crypto is interesting enough to steal the thunder from the foundational commodities known to finance.

References

1. SoFi Learn, Understanding The Different Types Of Cryptocurrency, June 19, 2019, https://www.sofi.com/learn/content/understanding-the-different-types-of-cryptocurrency/

2. Swati Goel, The Difference Between Fiat Money and Cryptocurrencies https://www.fxempire.com/education/article/the-difference-between-fiat-money-and-cryptocurrencies-520616#:~:text=Governments%20issue%20fiat%20currencies%2C%20which,regulated%20by%20the%20central%20bank.&text=Cryptocurrencies%2C%20on%20the%20other%20hand,control%20or%20influence%20their%20value.

3. ET, 5 Reasons Why You Should Go For Cryptocurrency, October 25, 2017, https://m.economictimes.com/industry/banking/finance/5-reasons-why-you-should-go-for-cryptocurrency/articleshow/61184608.cms

4. Shawn M. Carter, How Does Cryptocurrency Work?, February 24, 2020, https://www.foxbusiness.com/technology/how-does-cryptocurrency-work.

5. Cash Lambert, Cryptocurrency Basics: A Beginner's Guide, https://wealthfit.com/investing/cryptocurrency-basics/#:~:text=Cryptocurrency%20is%20digital%20money.,which%20you've%20likely%20seen.

CHAPTER 2

History Of Cryptocurrencies

Introduction

By now, we know what cryptocurrency is and how it is different from fiat currency. In the previous chapter, we discussed how crypto works, the different types available, and how to store, invest and trade cryptocurrency. This chapter deals with the history of crypto and discusses the future of Bitcoin and other such cryptocurrencies. The concept behind cryptocurrency is not as new as you may have thought. Before the invention of cryptocurrency, many people have made attempts at creating one. A notable example is Wei Dai, a computer engineer, who published a paper where he discussed "B-money" in 1998. He envisaged the idea of a digital currency that could be sent along with specific untraceable digital pseudonyms.

The same year saw Nick Szabo, a blockchain pioneer, create Bit Gold, an electronic currency system requiring users to complete a POW (Proof-of-Work) function with solutions put together cryptographically and published. The idea behind this concept was to overcome the inefficiencies within the traditional financial system. Though both B-money and Bit Gold never became mainstream, they were the primary inspiration behind the invention of Bitcoin.

2.1 The Rise of Bitcoin and Other Cryptocurrencies

The credit for inventing Bitcoin goes to Satoshi Nakamoto, whose identity remains a mystery to this day. In 2008, he published a white paper, "Bitcoin – A Peer-to-peer Electronic Cash System," describing the Bitcoin Blockchain Network functionality. The following year saw the Bitcoin software being made available to the public for mining.

Change from Mining to Trading

In the early days, Bitcoin was only mined and not traded. Hence, there was no possibility of assigning any monetary value to it. The first attempt at trading Bitcoin happened when a person named Laszlo Hanyecz decided to sell Bitcoin by swapping 10,000 Bitcoins for two pizzas. At that time, the transaction was worth around $41. Considering today's Bitcoin rate, it would be worth more than $400 million.

A Decentralized Economy

Bitcoin was the first decentralized cryptocurrency. As it started increasing in popularity, the idea of decentralized currency started to catch up. Other cryptocurrencies began appearing, with Namecoin and Litecoin being the first to emerge by 2011.

By July 2015, the Ethereum network was launched. Today, Ethereum has the distinction of being the second-largest cryptocurrency.

Though Bitcoin was a popular cryptocurrency, countries had not legalized it as a payment method. Japan was the first country which legalized Bitcoin in April 2017. Bitcoin had by then gained a reputation of being a safe mode of exchange. However, the year 2018 saw the hacking of Bitcoin exchanges in South Korea and Japan. As a result, Google and Facebook banned advertisements involving cryptocurrency. The year also saw tremendous fluctuation in Bitcoin's price as it fell below $4,000 in December 2018.

In 2020, Facebook displayed an interest in launching its cryptocurrency, Libra. Simultaneously, Bitcoin's value started spiking upwards.

2.2 The Devaluation of The Dollar

Between 2011 to 2013, Bitcoin grew in popularity to reach parity with the US Dollar in 2013. By May 2013, there were around ten crypto assets in the market. The same year saw Bitcoin touching $1000 for the first time. The price crashed to $300 and took more than two years to regain parity.

2017 was a turning point in Bitcoin's history as the price went above $10,000 for the first time. At the same time, we witnessed the Bitcoin cash hard fork that split Bitcoin into two derivative cryptocurrencies, Bitcoin Cash and BTC.

Reasons Behind Volatility of Bitcoin

Bitcoin pricing has always been volatile. The price of Bitcoin at the time of the first deal was $41 for 10,000 Bitcoins. Let us see the reason behind its volatility and how it has resulted in the US dollar's devaluation.

1. *The primary reason why Bitcoin price is increasing is the universal "demand-supply" relationship. As the supply goes down, the demand for the product goes up, and vice versa. Automatically, it changes the pricing. Bitcoin has a fixed supply as there is only 21 million Bitcoin that will ever exist. As this number is static, it makes Bitcoin scarce.*

2. *Secondly, the volatility is due to the 'Halving' process that is coded into Bitcoin. Bitcoin works on its traditional in-built Escrow mechanism where Bitcoin is released. Miners get Bitcoins as rewards for processing transactions. This reward halved every four years, with the last such halving happening in May 2020.*

3. *As a result of halving, Bitcoins' inflation rate is reduced to half, whereas its stock-to-flow ratio gets doubled. This process keeps repeating every four years until the entire stock of 21 million Bitcoins is released. In the third quarter of 2020, there are around 18.5 million Bitcoins in the world.*

4. *At this rate, the Bitcoin value could be somewhere around $100,000 in late 2021. Thus, you can see how the US dollar had devalued from 243.90 Bitcoins per dollar in 2009 to 0.000037 Bitcoins per dollar in 2020.*

2.3 The Future of Bitcoin and Other Cryptocurrencies

With Bitcoin prices crossing $27000 and close to $28,000 today, there are talks in the financial market that Bitcoin could successfully replace the US dollar as a global reserve currency. Though cryptocurrency has its utility, there are wild speculations regarding cryptocurrencies. The issue with crypto assets is that they can serve as a useful hedge or useful form of payment, but not both.

- ***New Payment Instruments:*** Deregulated and non-intermediated direct payment assets have demand. It explains why crypto assets such as "stablecoins" pegged to the US dollar have evoked interest in people. People trust transferring something linked to the dollar but with crypto-like features. However, trading in crypto assets has its risks. The stablecoin peg with the dollar can break. It is always a problem with pegged exchanges.

- ***Impending Regulations:*** Secondly, crypto-assets can become a part of the financial system, thereby attracting regulations. These crypto assets could lose the advantages they have over fiat currency and the traditional banking system under such circumstances.

- ***Rapid Clearing System:*** The formal banking sector could improve and move to a quick clearing system where you can transfer funds from one part of the world to the other and achieve a non-intermediated transfer within the central bank's accounting system.

Cryptocurrencies like Bitcoin or Ether are not pegged to national currencies. Therefore, they serve as useful hedges or speculation vehicles. However, they do not serve as dominant means of purchase. The reason is their volatility in their pricing. If the rates can go up rapidly, they can come down equally quickly. If you are using crypto for making small purchases, it should not be an issue, but it is risky in large transactions.

2.4 Cryptocurrency and The Global Regulatory Landscape

Cryptocurrency like Bitcoin prided itself that it is a deregulated means of exchange and saves time and costs. It is mainly true. It also explains why governments worldwide did not express any concern over the necessity to regulate cryptocurrency.

Things are changing today because there is increasing use of cryptocurrency as a means of exchange. Therefore, regulation seems to be a logical step forward. However, the law would entail compliance with KYC and AML. That can pose serious issues.

The US Financial Crimes Enforcement Network declared in 2013 that cryptocurrencies do not qualify as legal tender in any jurisdiction. However, Japan became the first country to introduce regulations in 2016 after the series of cyberattacks against crypto exchanges in South Korea and Japan. It included the massive theft of 850 thousand Bitcoins against Mt. Gox. Nevertheless, many governments across the globe still agreed that there was no need to regulate cryptocurrencies.

Even as late as 2018, The Financial Stability Board contended that cryptocurrencies did not pose any risks to global financial stability. However, many countries have prohibited dealing/

trading in cryptocurrencies. These countries include China, Bangladesh, Algeria, Bolivia, Ecuador, Colombia, Egypt, Iran, Indonesia, Nepal, Taiwan, and many more. At the same time, countries like Switzerland, Mexico, Canada, Japan, and the US have regulated cryptocurrencies. Countries like Australia, Singapore, the European Union, the UK, Russia, and South Korea consider cryptocurrencies as grey areas that require regulations.

The divided opinion among countries concerning dealing in cryptocurrencies leads us to discuss the fears governments have with crypto trading. One such fear is that malicious actors can use cryptocurrencies to launder money or even finance terrorism. All ransomware dealings happen in cryptocurrencies. The infamous Silk Road Case shows how drugs, guns, and passports got exchanged for Bitcoins in the early days of cryptocurrency. However, many experts contend that fiat currencies could also get used for the same purposes.

Crypto advocates highlight its benefits by stating that cryptocurrency and its underlying technology, Blockchain, have proved to be ideal solutions as they create a secure and transparent chain to reduce the risk of fraud across different sectors drastically.

2.4.1 The Country-Wise Position Regarding Cryptocurrency Regulations

This table details the regulation and upcoming legislation status of cryptocurrencies of some of the major countries.

Country	Cryptocurrencies	Crypto Exchanges	Exchanges	Future Legislation
The US	Not legal tender	Legal, with regulation varying from state to state	Uncertainty	In the pipeline
Canada	Not legal tender	Legal must register with FinTRAC after June 01, 2020	Inconsistent	No signs of any future legislations
Singapore	Not legal tender	Legal requires registration with the Monetary Authority of Singapore.	Under the regulatory authority of MAS	Could come up from time to time
Australia	Legal, treated as property.	Legal, registering with AUSTRAC, is compulsory.	Regulated	Tightening of norms expected
Japan	Legal and treated as property.	Legal, but must register with the Financial Services Agency.	Legal	Ongoing
South Korea	Not legal tender	Legal, resister with FSS	Strict Government regulations	Ongoing
China	Not legal tender	Illegal	Considering a ban	No indication of the lifting of the ban
India	Not legal tender	Effectively illegal, but regulations are being considered.	Reserve Bank of India had banned banks from dealing in cryptocurrency, but the Supreme Court ruled the ban unconstitutional.	Revisiting policies of regulating cryptocurrencies
The UK	Not legal tender	Legal, register with FCA.	Registering with FCA is mandatory.	No specific legislation on the horizon
Switzerland	Legal, accepted as payments	Legal, regulated by SFTA	Must obtain a license from FINMA	Considering encouraging innovations in Blockchain technology.
The EU	Legal, member states may not introduce their cryptocurrencies.	Regulations vary from one member-state to another.	It depends on each member-state	Exploring further cryptocurrency regulations

Final Words

This chapter dealt with the history of Bitcoin and other cryptocurrencies. It also explored areas regarding legislation and regulations imposed by countries globally. Though there is divided opinion regarding crypto usage among nations, there is little doubt that cryptocurrency will be the way forward and poised to revolutionize the international financial market.

References

1. *Crypto Throughout The Years a Short (Ish) History – Part 1* Gatehub. May 19, 2020. https://gatehub.net/blog/history-of-cryptocurrency-part-1/

2. *A Brief History on Bitcoin and Cryptocurrencies. Ledger Academy.* October 23, 2019. https://www.ledger.com/academy/crypto/a-brief-history-on-bitcoin-cryptocurrencies

3. *Bernard Marr. A Short History Of Bitcoin And Crypto Currency Everyone Should Read. Forbes.* https://www.forbes.com/sites/bernardmarr/2017/12/06/a-short-history-of-bitcoin-and-crypto-currency-everyone-should-read/?sh=5a9be3103f27

4. *The History of Cryptocurrency. Bright Finance.* https://www.brightfinance.co/the-history-of-cryptocurrency.html

5. *Raynor de Best, Number of Bitcoins in circulation 2012-2020. Statista. November 23, 2020.* https://www.statista.com/statistics/247280/number-of-bitcoins-in-circulation/#:~:text=In%20the%20third%20quarter%20of,Bitcoin%20is%20a%20virtual%20currency.

6. *Luke Conway. Price Rising? Investopedia. November 05, 2020.* https://www.investopedia.com/tech/cryptocurrency-this-week/

7. *Tyler Cowen. Cryptocurrency is not necessarily the future. Bloomberg Opinion. December 29, 2020.* https://www.bloomberg.com/opinion/articles/2020-12-29/bitcoin-hits-new-high-but-cryptocurrency-s-future-is-uncertain

8. *The Evolution of Cryptocurrency and Regulations since the inception of Bitcoin. Electroneum.* https://news.electroneum.com/the-evolution-of-cryptocurrency-and-regulations-since-the-inception-of-bitcoin

9. *Cryptocurrency Regulations Around The World. Comply Advantage.* https://complyadvantage.com/blog/cryptocurrency-regulations-around-world/#

CHAPTER 3

Trading Bitcoin

Introduction

Bitcoin trading involves the ability of investors to speculate on the crypto price movements. Traders are making the most out of the volatility of Bitcoin by deploying unique trading strategies. This chapter discusses details on trading Bitcoin. Bitcoin trading refers to the purchasing and selling of this popular cryptocurrency. With its volatility and traded volume, Bitcoin trading offers lucrative avenues to investors. Trading in Bitcoins involves speculations on how the prices of these cryptocurrencies move. The process involves purchasing Bitcoins through an exchange and selling them off on the same day or holding them for the long term. Traditionally, the mechanism is similar to stock market trading. However, trading in cryptocurrencies has its benefits, which makes it popular among investors. Thanks to the security and volatility of Bitcoin, investors can reap huge profits.

3.1 Different Options Available for Trading Bitcoin

Like stock trading, Bitcoin trading also involves different approaches. These include:

- ***Trading in Bitcoin derivatives***: Derivatives refer to tradable contracts or securities which use an underlying asset to derive their value. The underlying asset is mostly Bitcoin (BTC) for cryptocurrency derivatives.
- ***Trading in Bitcoin options***: Bitcoin options trade similar to other call or put options where investors pay a premium for the right (not obligation) to sell or buy an agreed amount of the cryptocurrency on an agreed date.
- ***Trading in Bitcoin futures***: Investors can gain exposure to Bitcoin without holding the underlying cryptocurrency through Bitcoin futures. They allow investors to speculate on Bitcoin's future price.

3.2 Ways to Trade Bitcoin

One can deploy different types of Bitcoin trading strategies to generate profit. In this case, a strategy refers to a methodology for trading Bitcoins covering the prices at which a trader enters and exits the market. Here are some of the most popular strategies to trade Bitcoins.

- ***Holding Bitcoins (HODL):*** Holding Bitcoins happens to be one of the most effective and robust trading strategies. This strategy involves maintaining a position on Bitcoin on a long-term basis. However, users should note that Bitcoins are incredibly volatile. Inexperience and drastic drop in prices can lead to losses. Holding Bitcoins often pay off for investors as their prices soar over the long-term. For instance, in 2018, the Bitcoin value soared to its multi-year peak. Therefore, investors can enjoy a long-term benefil if thoy invest when the prices are still low. Unless you have a plan for risk management in place, you should refrain from deploying this technique.

- ***Hedging Bitcoin*:** Investors who own Bitcoin can hedge their risk if they think that the market is going on a short-decline. Hedging refers to the practice of indulging in strategic trades to decrease the risks to existing positions. To hedge Bitcoin's existing holding, investors can open a position to short Bitcoin, meaning they can sell it at the current market price, expecting it to decline. In case the market falls, they can repurchase it for a lower price and make a profit.

- ***Trend Trading Bitcoin*:** A trending market reaches extreme highs and lows consistently. This strategy is suitable for varying timeframes in which investors hold their position open as long as they believe the trend will continue – hours, days, weeks, or months.

- ***Bitcoin AI Strategies*:** It involves computers discovering strategies for trading Bitcoin on their own through evolutionary learning techniques. All markets have inefficiencies that can be exploited for consistent gains. A.I can discover patterns in everything from price data to Bitcoin premium (see Chapter 5).

3.3 Costs Associated With Moving In And Out Of Bitcoin

Like stock trading, Bitcoin trading also involves individual costs and fees. Each time you make a transaction in a cryptocurrency, you would have to shell out a small percentage in terms of fee. Before you start investing in Bitcoins, you should be aware of all the associated costs.

- ***Trading Fee*:** This is the charge levied for buying or selling Bitcoins on the cryptocurrency exchange.

- ***Maker Fee*:** Maker fee refers to the charge levied for placing an order that does not execute immediately. For example, the exchanges charge a maker fee on limit orders.

- ***Taker Fee*:** The crypto exchanges charge a fee when an investor places a Bitcoin order, and it executes immediately. Market orders involve a taker fee, as you make the transaction regardless of the price.

- ***Miner Fee:*** As you are aware that Bitcoins have to be mined. The miner fee is levied to pay the computers and people involved in mining them.

In this context, you should note that the charges are different for each cryptocurrency exchange. While some of them levy a commission for each transaction, others may charge the taker or maker fees. It mainly depends on the transaction's nature.

3.4 Risks and Security Issues Associated With Cryptocurrency Trading

Now, let us find out the risks involved in trading in Bitcoins.

- ***Technological Risks:*** The development of technical aspects often takes place uncontrollably and quickly. The users are aware of the fact that the number of competitors for Bitcoin continues to rise daily. In this case, the advantage is the massive brand awareness and the availability of large venture capitals. However, a technological risk persists, as a more advanced cryptocurrency may appear in the market. It indicates that investors might fail to notice the movement when Bitcoins lose their value.

- ***Operational Risk:*** In a centralized clearinghouse, the transactions are validated, and they can be reversed if necessary. However, cryptocurrency transactions cannot get reversed. Another threat to the investors is the lack of permanence since the accounts are secured cryptographically. In case the keys to a particular account are stolen or lost, the investors would lose access to the monies in the account. Eventually, it might get deleted forever for the actual owner.

- ***Market Risks:*** Liquidity concerns remain an issue in Bitcoin trading, as the amount of currency available in the market is finite. Bitcoin trades take place only on demand, and therefore, the risks in the market are idiosyncratic. Chances of manipulation in the market cannot be ruled out, as Bitcoin ownerships are limited. The lack of alternatives and limited acceptance makes Bitcoin more volatile, as compared to physical currencies. The impact of hoarding and speculative demand remains some of the other concerns that investors should understand.

- ***Compliance and Regulatory Risk:*** In the future, certain countries may prevent their citizens from using Bitcoins. The authorities might also consider these transactions to break regulations formulated to avoid money laundering. The senders, processors, and receivers of Bitcoin would not enjoy a single or uniform approach considering the significant number of investors and complexity regarding its decentralized nature. Eventually, investors may face regulatory or compliance issues while they hold Bitcoins.

- ***Risk During Experimental Phase:*** Cryptocurrency itself is an innovative concept. Investors hardly have access to any historical data to assess the degree to which they can rely on it. Besides, Bitcoin is still developing like any other cryptocurrency. Therefore, one cannot rule out the uncertainty factor altogether. With experimental technologies, something unexpected can happen at any time. However, Bitcoin is considerably less experimental than other cryptocurrencies as it is used most actively.

3.5 Risk Management in Crypto Trading

Trading in cryptocurrencies comes with some inherent risks. Therefore, derivative traders in cryptocurrency need to incorporate specific risk-management strategies. Particularly, traders need to address the volatile nature of this digital currency. The daily range is too extensive on average. In the absence of any risk-management strategy, you might end up with a loss.

Here are specific tactics traders need to deploy for risk-management.

- ***Deploying Stop Losses And Making Partial Profits:*** To mitigate the extent of loss, traders need to use strict stop losses. Simultaneously, one needs to be strategic while closing the profit-making positions in Bitcoins before the trend starts changing. During abrupt changes in the market, these two strategies can help you optimize your profits.
- While trading Bitcoins, traders often experience the movement of the market in the opposite direction. In case you have open trades, you would lose all your money if you have not used a stop loss. On the other hand, if the trade takes place in the same direction in which the price moves, you need to book profit before a reverse trend takes over the price movement. In both cases, one must make sure not to close the trade too early.
- ***Don't Indulge In Revenge Trading:*** Traders who are just starting Bitcoin trading often indulge in revenge trading. After incurring a loss, emotions begin to take over the rational mind. Eventually, they place another trade to recover the loss immediately. However, they hardly realize that they did not include this trade in their strategy, and they are likely to suffer another loss. Thus, they lose a significant part of their capital in the process.

- ***Run Few Trades Or Contracts At A Time:*** During Bitcoin trading, it is recommended to open a few trades or contract at any given point in time. It ensures that you would have an adequate margin to hold these trades for a long time, if necessary. The margin amount will be more considerable if you run a few trades. To make the expected profits, you can hold the trade for a long. Your Bitcoin account balance largely determines the number of trades. In case you have a low balance, do not indulge in more than two open trades at a time.

- ***Adhere To Your Trading Strategy:*** Being a trader, you need to accept a few losing Bitcoin trades at a time. No strategy is flawless. Besides, you might experience inconsistency in profit-making. However, if you realize that the overall returns are coming in the form of profits, you need not worry about losses. Regardless of the outcome of a particular moment, you need to adhere to your trading strategy. Traders should abandon a strategy only when the losses exceed their profits. Investors just getting started with Bitcoin trading must use a demo account at the outset.

Final Words

To wrap up, it is evident that Bitcoin trading can be highly lucrative for disciplined traders. With the right strategy in place, Bitcoin trading carries more profit-making potential than forex or stock markets. For investors, it is imperative to use a reliable platform or exchange to stay safe from fraud. Trading in Bitcoins can be highly profitable for non-conventional traders, who apply proper stop losses and adequate risk management.

References

1. *Nicole Deiker. How to Trade and Invest in Bitcoin & Other Cryptocurrencies. Retrieved from:* https://www.mybanktracker.com/blog/investing/fees-trade-invest-Bitcoin-cryptocurrencies-276084

2. *IG. How to trade Bitcoin. Retrieved from:* https://www.ig.com/en/Bitcoin-btc/how-to-trade-Bitcoin#:~:text=There%20are%20a%20few%20different,Crypto%2010%20index

3. *Bitwala Trading. What is Bitcoin trading? Retrieved from:* https://www.bitwala.com/academy/what-is-Bitcoin-trading/

4. *IG. Best Bitcoin trading strategies and tips. Retrieved from:* https://www.ig.com/en/trading-strategies/best-Bitcoin-trading-strategies-and-tips-190813

5. *IG. What is cryptocurrency trading, and how does it work? Retrieved from:* https://www.ig.com/en/cryptocurrency-trading/what-is-cryptocurrency-trading-how-does-it-work

6. *John Thackeray. 5 Inherent Risks of Cryptocurrency. Retrieved from:* https://daily.financialexecutives.org/FEI-Daily/July-2018/5-Inherent-Risks-of-Cryptocurrency.aspx

7. *FinExtra. The Main Risks Of Investing In Bitcoin. Retrieved from:* https://www.finextra.com/blogposting/17390/the-main-risks-of-investing-in-Bitcoin

8. *Nicolas Tang. Risk Management in Cryptocurrency Derivatives Trading. Retrieved from:* https://phemex.com/blogs/risk-management-in-cryptocurrency-derivatives-trading

CHAPTER 4

Best Trading Strategies For Retail Traders

Introduction

Bitcoin trading attracts large masses of retail traders. It is so because Bitcoin is trading at all-time highs after the 2017 mainstream boom. High volatility, high leverage, low margins, no KYC requirements, etc., make Bitcoin trading an attractive investment for new traders. This chapter discusses the various Bitcoin trading strategies which they can employ.

Many people have adopted trading as a passive income source for the ease of buying and selling and the remuneration one can get in the long-run. Moreover, since the internet is equipped with resources to teach beginners how to trade in the cryptocurrency segment profitably, many tend to trade and earn well. But it is not without its demerits - even though a beginner trader might make money for a week or so, they tend to make huge losses due to a lack of systematic trading methodologies.

While trading, it is essential to understand that the economy goes through various cycles of expansion and depressions, and the price of multiple cryptocurrencies varies accordingly. Therefore, a single strategy will not be successful during such fluctuations. Thus, experienced and successful traders usually employ a hybrid of techniques to make money consistently in varying economic conditions. Most importantly, it is evident that traders will lose more than they earn without patience and planning.

4.1 Before You Start Trading

Traders need to ensure various etiquettes before initiating the buying and selling of cryptocurrency like trading stocks. These provide the security of the asset and monetary benefits in the long run. Multiple measures that retail investors must follow before trading cryptocurrencies are discussed below.

- ***Choose a reputed exchange:*** Cryptocurrencies are traded through various exchanges like bonds and stocks are traded. Even though there might be fees, it is safer to trade through reputed and verified exchanges.
- ***Secure email address:*** Cryptocurrencies such as bitcoins and ethereum will be linked to a wallet connected to your email address. If the email account is compromised, the malicious actors can easily trade and cause harm to your trading account.
- ***Learn safe wallet procedures:*** Wallets where cryptocurrencies are held, should be secured using maximum security, such as using private keys and double encryption to avoid getting hacked by intruders.
- ***Identify the taxes to be paid and brokerage charges:*** Since cryptocurrencies are traded through regulated exchanges, traders are liable to pay taxes annually based on their operation area, and the amount of profit retail traders get yearly. Similarly, brokerage charges are small charges payable to the broker under which retail traders buy and sell cryptocurrencies.
- ***Learn technical analysis:*** Technical analysis (with a computer) uncovers various price patterns and market trends, useful for short-term retail traders. Using a computer to confirm these patterns is a must in order to have an unbiased opinion -- especially when it comes to price patterns. People see patterns in everything from clouds to charts (known as pareidolia).
- ***Find a mentor:*** Finding an experienced and successful mentor can help avoid huge losses during the initial stages, and traders could learn a lot of useful strategies from professional traders.

- ***Trade only a small percentage of total assets:*** It is recommended not to use more than 50% of real liquid assets in the market during the initial stages. Moreover, always have enough funds for emergencies such as health conditions or paying off loans and mortgages. The amount of trading should not come from your daily expenses.

- ***Beware of scams:*** The cryptocurrency market is notorious for various scams, and these fraudsters usually demand bitcoins from victims claiming they will double your amount within a week. Always trade through legitimate brokers and exchanges and never share your account credentials with third party services.

- ***Always avoid being greedy:*** In the initial stages, traders may witness huge profits all of a sudden, and the market lures them to invest more. But beware, losses can occur in the same manner and cause substantial financial burden if there is no stop loss assigned.

- ***Understand the underlying technology:*** Cryptocurrencies are based on blockchain. Digging deep into blockchain and other related technologies will help traders understand how market sentiments react according to the technology employed by each cryptocurrency. Analyzing the technology used is fundamental for cryptocurrencies since they are based on blockchain and decentralized networking methodologies.

4.2 How To Trade Effectively, And Make Profits?

Once a retail trader has followed the initial steps and opened a wallet for storing cryptocurrencies, it is time to buy low and sell high. Profits drive traders to trade more, and hence, to ensure profit, it is necessary to apply a trial and error technique on various trading strategies. The following rules can help traders to minimize losses and earn higher profits in the long-run.

- ***Bitcoin spread:*** Since bitcoins have a considerable share in the entire crypto market, prices of other cryptocurrencies tend to rise and fall as bitcoin and their bretheren prices vary. Cryptocurrencies are thus closely linked to Bitcoin. When the relative prices between cryptos becomes large, prices tend to revert to the mean.

- ***Set stop loss:*** For active retail traders setting a stop loss while taking on positions can avoid huge losses due to sudden price crashes in cryptocurrencies.

- ***Diversification:*** If a trader owns various cryptocurrencies, it will ensure that price fall in one asset does not entirely affect the entire portfolio. Thus, the risk is diversified by possessing multiple financial instruments.

- ***Perform fundamental analysis:*** Fundamental analysis will provide an outlook on how the cryptocurrency price will be after a few years and helps analysts understand which asset to place maximum and minimum capital and thus undertake strategic investment decisions.

- ***Follow planned entry and exit techniques:*** Trades should happen in a pre-planned manner, and entry price and exit price should get determined even before buying an asset. It ensures that the trader follows a planned and algorithmic strategy, which will be useful in the long-run.

4.3 How To Make Money When Bitcoin Declines?

Like traders making money when cryptocurrency prices rise by buying low and selling high, traders can short the cryptocurrency when bitcoin prices tend to go down. Traders, either retail or institutional, can take short positions in bitcoins through the following methods.

- ***Margin trading:*** In this method, a trader can borrow money for buying cryptocurrency from brokerages if the trader has some particular percentage of money in their wallet. Since the extra fund is borrowed, the risk involved is relatively high.
- ***Bitcoin futures:*** Futures trading is a derivatives segment trading in cryptocurrency. The trader agrees to pay for the bitcoin after a predefined period and is ready to transfer the cryptocurrency to his wallet later.
- ***Exchange-Traded Fund (ETF):*** At this time there is not a Bitcoin Exchange Traded Fund.
- ***Binary contract:*** Binary contract helps traders to earn from the percentage of the price drop of bitcoin. A retail trader sells the cryptocurrency before buying and can buy at a lower price on the same day and claim his profit.
- ***Short-Selling the Bitcoin Assets:*** This may not appeal to all investors. It involves short selling Bitcoins directly or selling the tokens when they anticipate that the market is going down, waiting until the price drops, and buying back the tokens again.

4.4 How To Trade Securely: What Not To Do?

Trading is a continuous game, and a handful of profit earning trades does not define success rate. In addition to profits, there are various aspects to it. Therefore to securely trade and book profits, it is suggested to follow the below-discussed measures while trading.

- ***Never trade without a plan:*** Planning is a crucial step in trading, and stop-loss, entry, and exit strategies should be predetermined before taking positions. Before taking each trade, always analyze the trade's objective and how you will complete it.

- ***Never buy the hype:*** Cryptocurrencies, which are already noted as the next multi-bagger which have inflated prices and buying at the peak, will most probably welcome a losing position into a trader's portfolio. Moreover, some traders tend to spread fake news and benefit from price hikes during the same and existing from the position before others discover that it was a hoax.

- ***Never buy cryptocurrencies that are not researched well:*** The essential character of a trader should be the effort he puts forward into research, such as performing fundamental and technical analysis, determining the market trend, and analyzing the scope of the cryptocurrency. If a trader trades without relevant research, it equates to gambling, and they can quickly lose money.
- ***Never trade without an exit plan:*** Trading, specifically intraday trading, should never be executed without assigning stop loss. Moreover, setting a stop loss will ensure that positions are automatically squared off during abrupt market crashes that happen overnight.

4.5 Crypto CFD - Good or Bad?

Crypto CFD trading or contract for difference trading is an agreement between client and broker without actually taking a market position. Crypto CFD is advantageous in various ways since it offers overseas trading and allows short selling and taking leverages for trading in cryptocurrencies. Moreover, since the trading activity tends to increase annually and volatility in the market increases in parallel. Such an increase in volatility creates an urgency to implement fast transactions since a delay in executing trade might lead to some losses in the trader's portfolio.

Above all primary advantage of crypto CFD such as bitcoin CFD is that they are regulated. Unlike traditional bitcoin trading, they are not unregulated, and traders and investors have a complaint portal to register complaints against frauds and other criminal activities occurring in the digital world. CFD trading has been evolving as an alternative to investment and as a hedging instrument in the market.

Final Words

Traders with access to supercomputers do not control the cryptocurrency markets. Hence there is a more level-playing field in the crypto market for retail traders. Additionally, with the market cap of cryptocurrencies crossing $700 Billion, it presents a massive opportunity for early investors. While some traders might complain about the fluctuating markets, if one knows how to trade them with innovative strategies as mentioned above, they can rake in considerable amounts of profit!

References

1. *TechniTrader "Trading Strategies." ,* https://technitrader.com/trader-wiki/trading-strategies/

2. *Rouchy, Tony, "10 Things You Need to Do before Trading Cryptocurrency." 22 Feb 2019,* https://medium.com/@rouchy/10-things-you-need-to-do-before-trading-cryptocurrency-69248301a3a1.

3. *Subham. "How to Make Money When the Bitcoin Price Drops.", 26 Sept 2019,* http://www.techtipsapp.com/make-money-when-bitcoin-price-drops.

4. *Cryptohopper, "The Do's & Don'ts Of Crypto Trading."* https://www.cryptohopper.com/blog/254-the-do-s-don-ts-of-crypto-trading.

5. *Pedro Febrero, "What Is Cryptocurrency CFD Trading?" 7 June 2019,* https://coinrivet.com/guides/what-is-cryptocurrency-trading/what-is-cryptocurrency-cfd-trading/

CHAPTER 5

Trading Like Stocks: Trading With GBTC (Grayscale Bitcoin Trust)

Introduction

In the previous chapters, we have seen various Bitcoin trading strategies. We know that trading in Bitcoin requires setting up and maintaining a separate account and paying high trading fees. This chapter looks at trading Bitcoin like stocks through the Grayscale Bitcoin Trust (GBTC). It is the fastest way investors can get crypto exposure without buying Bitcoin.

The Grayscale Bitcoin Trust (GBTC) is an investment product for holding digital currency, enabling individual investors to buy and sell it in their brokerage accounts. It registered its shares with the SEC and became an SEC reporting company on January 21, 2020. Thus, the Commission designated the trust as the first investment vehicle for digital currency, which attained a reporting company's status by the SEC. According to the SEC rules, accredited investors can now purchase shares in GBTC's private placement and grab an early liquidity opportunity since the private placement shares' statutory holding period will be reduced from twelve to six months.

5.1 Why Should You Trust GBTC?

GBTC is an investment vehicle trading over-the-counter, which means that investors can buy and sell it like any other U.S. security. For example, investors can trade GBTC through brokerage firms or tax-advantaged accounts like 401(k)s or IRAs. The advantages of choosing GBTC over Bitcoin are:

- ***Excellent for casual buyers***: Generally speaking, GBTC is an excellent buy for everyday investors looking to trade Bitcoin but do not want to trade crypto on an exchange. Because of its limited supply and high demand, GBTC trades at a high premium (the difference between the value of the holdings and the market price).
- ***Higher profits:*** Because of its high premium and the Bitcoin market's volatility, investing in GBTC can become off-putting sometimes, but it can offer higher profits than Bitcoin.
- ***No cap on spending:*** Investors need to maintain a digital wallet and upload personal information on a crypto exchange, allowing them to convert their fiat currency (USD) into cryptocurrency (Bitcoin, Ethereum, etc.). It is a slow process, and most crypto exchanges limit the investor's spending capacity. However, investors who trade in GBTC are not subject to a cap on their spending.
- ***A cheaper option than brokerage:*** Traders often need to pay fees when trading Bitcoin (for USD or other currencies). The costs can exceed the 2% required to be paid to the trust (paid from the trust and not paid per transaction). For investors whose fees are low or who are not paying it with their brokerage, GBTC can be a cheaper option for them.
- ***Robust performance history:*** Even with the high premium, GBTC has a good performance history. Although occasionally inconsistent, it can prove to be profitable for investors having the right timing.
- ***Significant profits:*** The premium sometimes allows investors to rake in considerable profits. If they buy during a bear run

of the premium and wait till it turns bullish, investors can outperform Bitcoin by trading in GBTC. (Cautionary note: The above suggestion can work the other way too!)

- ***A tax-friendly exposure:*** There are various tax advantages when trading on publicly-traded Bitcoin trusts like GBTC. Many Roth IRA, IRA, and other investor and brokerage accounts do not give tax rebates for Bitcoin investments, but they give them on publicly traded trusts. Thus, GBTC offers traders a tax-friendly exposure to Bitcoin.

5.2 The Best Way to Trade Securely and Make Profit

Here are a few safety tips if you are trading in GBTC:

- ***Know what you are buying***: One share of GBTC= 0.001 BTC (approx.). The last trade of GBTC was at $44.96 for 1 BTC. If we do the necessary calculation and compare it with the current market rate of Bitcoin at $38,789, it means that owning 1 Bitcoin worth of GBTC will cost (44.96 X 1000 = $44,960), or a premium of approx. 15% . You must decide if the value of holding Bitcoin in your IRA justifies the cost of paying a 15% premium.
- ***Avoid using large market orders***: GBTC trades over the counter (OTC) at this time. While GBTC has dramatically increased in liquidity, entering large market orders can generate slippage (bad fills). It's often better to break orders into multiple orders less than 500 shares.

- ***Prefer using limit orders***: The advantage of using *limit* orders is that they guarantee the maximum price an investor is willing to pay. Thus, it protects you from bad execution risk or paying more than what you are ready for.

5.3 Is There a Bitcoin Fund You Can Invest in?

Currently, there are no Bitcoin ETFs or mutual funds. The GBTC helps fill the gap until a bitcoin-based ETF gets approved by the SEC. It allows investors to gain exposure to Bitcoin without the need to buy it. GBTC operates like a fund where its team of highly qualified professionals pools the investors' money into buying Bitcoin. Some investors are wary of investing in the GBTC because it usually trades at a premium 20-30% above the Bitcoin value. GBTC offers them a product that tracks the value of Bitcoin (one-tenth of Bitcoin's worth) for those looking to invest. For example, if the Bitcoin value is $2,000, each of GBTC's share will have a net asset value of $200. This value includes the costs, including the 2% fee, which GBTC charges for maintenance.

5.4 How To Trade Using GBTC?

GBTC trades publicly on the over the counter market or OCTQX and falls under the **Alternative Reporting Standard** for organizations that do not have to register with the SEC (Securities and Exchange Commission). The success of GBTC shares mirrors that of Bitcoin because it derives its value solely from Bitcoin. The fund has $13.1 billion in AUM (Assets Under Management), and its average daily trading volume is over 15.95 million shares a day. Grayscale stores its assets with Coinbase Custody Trust Company, LLC, which protects it with robust cryptographic security. The expense ratio is 2% to cover the cost of safekeeping and maintenance. Investors must know how the Grayscale Bitcoin Trust works to implement strategies and trade with GBTC to earn a profit.

- Grayscale invites wealthy investors to create a private pool and pledge money to the fund. It then uses this money to buy large amounts of Bitcoin. By wealthy investors, it is meant that GBTC allows investments only from accredited investors who have a net worth of over $1 million or who have total earnings of over $200,000 (or $300,000 jointly). The minimum investment amount is $50,000.

- Then Grayscale will list the fund on public stock exchanges allowing traders to buy and sell shares in it. Since GBTC is one of the largest Bitcoin holders globally, it provides investors direct access to the cryptocurrency. Its sole asset is Bitcoin, other than the cash for overhead and administrative expenses. Hence, it offers a lucrative opportunity to have exposure to Bitcoin, especially the investors who have a brokerage account but not the time to understand the digital asset ecosystem's working.
- Next, shares in the fund track the price movement of Bitcoin. Similar to the SPDR Shares ETF (GLD) tracking gold's underlying value, GBTC tracks the underlying value of Bitcoin. Shares of GBTC can either trade at a premium or a discount compared to the price of Bitcoin. Historically, the shares of GBTC have traded at a premium, which benefits Grayscale and not the investors. But, a rapid increase in trading will help investors book higher profits in the future.

5.4.1 Investment Strategies

- Investors can use the private placement route to buy GBTC issuances cheaper because they are buying at the daily NAV (Net Asset Value).
- Another option is going for the publicly quoted GBTC trades, which come at a premium, driven by market forces and client demand.
- The GBTC trade price is higher in the secondary market than the NAV of the underlying Bitcoin holdings because it lacks the redemption feature.
- When the bitcoin is looking bullish, a profitable strategy could be buying GBTC when the premium to NAV hits bottom and sell when it reaches the highs. It can prove to be more beneficial than going for Bitcoins.
- For the past five years, the GBTC average premium was 39%, which traded between 132% to -0.34% but never at par or 0%. Thus, it presents a unique opportunity for hedge funds to use the arbitrage opportunity. They buy the primary issues at par and sell them in the secondary market at a premium six months later.

5.5 GBTC's Unique AI-based Trading Algorithm

Individuals and institutions who need to sell or purchase digital assets often have to transact through unfamiliar intermediaries or exchanges that are insecure or unregulated. These require them to transmit funds through uncomfortable jurisdictions. Additionally, storing digital assets come with additional risks because the private keys, which are the equivalent of passwords concerning digital assets, are susceptible to theft or loss. The private keys provide access to the investor's digital assets, and losing them exposes their confidential information to a partial or total loss. Thus, the investors are left with limited or no means to regain access to their digital assets.

In contrast, when investors choose to invest their shares in GBTC, they can avoid the challenges of purchasing, storing, and transferring the digital assets personally. The responsibility passes on to Grayscale and its service providers, who maintain the assets' security and provide investors with exposure to their assets' price performance. GBTC's unique AI-based trading algorithm and robust security procedures maintain the safety of digital assets. The security procedures include cold or offline storage, 2-step verification, multiple encrypted usernames, passwords, and private keys. GBTC's assets are stored in cold or offline storage with the Coinbase Custody Trust Company, LLC. It acts as a custodian of the digital assets and is a qualified custodian satisfying Rule 206(4)-2(d)(6) of the Investment Advisers Act of 1940, as amended. The LLC is also a fiduciary under the New York Banking Law. Experienced technology, security, legal, and financial professionals design these system protocols overseen by a financial operations team. Friedman LLP audits the financial statements of each Grayscale product.

Bonus Ticket

As a special bonus for readers, I'd like to invite you to a complimentary masterclass for those looking to generate retirement income with Bitcoin.

Registration is about to close. Act now before this opportunity passes you by.

"How to Buy Bitcoin at Massive Discounts with "Code-V Orders"

>> Register Now <<

References

- Nathan Reiff, "What Is the Grayscale Bitcoin Trust?" January 21, 2020, https://www.investopedia.com/news/why-buy-expensive-bitcoin-etf-instead-actual-bitcoin/
- Cryptocurrency Facts, "Understanding The Bitcoin Investment Trust (GBTC)" https://cryptocurrencyfacts.com/understanding-the-bitcoin-investment-trust-gbtc/
- Matthew Frankel, "Should You Use Grayscale Bitcoin Trust to Invest in Cryptocurrency?" December 16, 2020 https://www.fool.com/investing/2020/12/16/should-you-use-grayscale-bitcoin-trust-to-invest-i/
- Grayscale, "FAQs" https://grayscale.co/faq/

DAN MURPHY

Made in United States
Orlando, FL
05 November 2026

90200313R10036